The Gut Check cookbook

200+ Delicious Recipes you'll Want to Eat to Help you Unleash the Power of Your Microbiome to Reverse Disease and Heal your Gut

By Cherri J. Diaz

The Gut Check cookbook: 200+ Delicious Recipes you'll Want to Eat to Help you Unleash the Power of Your Microbiome to Reverse Disease and Heal your Gut

© [2024], [Cherri J. Diaz]

Table of Content

Gut-Healing Grains and Vegetables 75

One-Pot Meals for Easy Digestion 81

Introduction

Welcome and Overview

Welcome to "Gut Check: Unleashing the Power of Your Microbiome Cookbook." In these pages, embark on a journey to transform your health by harnessing the incredible potential of your gut. This cookbook is not just a collection of recipes; it's a guide to understanding and embracing the profound connection between what you eat and the well-being of your microbiome.

Our gut is a bustling ecosystem of trillions of microorganisms, influencing not only our digestion but also playing a pivotal role in our overall health. "Gut Check" is here to empower you with over 100 delicious recipes carefully crafted to nourish and heal. Whether you're aiming to reverse disease, enhance your energy levels, or simply adopt a gut-friendly lifestyle, this cookbook is your companion on the path to vibrant health.

As you explore these pages, you'll discover the importance of probiotic and prebiotic-rich foods, the significance of fiber, and the transformative impact of mindful eating. Each recipe is a testament to the idea that eating for gut health can be a delight to the senses. From breakfast to dinner, snacks to desserts, and beverages to cleansing recipes, "Gut Check" offers a diverse array of culinary creations that not only taste good but also prioritize your well-being.

So, let the journey to a healthier, happier gut begin. Embrace the joy of cooking, savor the flavors that support your microbiome, and unlock the potential to reverse disease while relishing every bite. "Gut Check" is more than a cookbook; it's your guide to unleashing the power within, fostering a harmonious relationship with your microbiome, and taking charge of your gut health. Let the culinary adventure for a healthier you commence!

The Importance of a Healthy Gut: Understanding the Microbiome

The human body is a marvel of complexity, and within it exists a thriving ecosystem that plays a critical role in our overall well-being—the gut microbiome. Comprising trillions of microorganisms, including bacteria, viruses, fungi, and other microbes, the microbiome is a dynamic community residing in the gastrointestinal tract. The health of this intricate system profoundly influences not only our digestion but also our immune function, mental well-being, and susceptibility to various diseases. In this exploration, we delve into the significance of maintaining a healthy gut and understanding the microbiome's intricate dance within.

The Gut Microbiome: An Overview

The gut microbiome is a vast and diverse collection of microorganisms that reside in the digestive tract, primarily in the large intestine. This bustling community is made up of bacteria, viruses, fungi, and other microbes, collectively forming a symbiotic relationship with the human body. In fact, the number of microbial cells in the human body is estimated to be comparable to, if not greater than, the number of human cells. This realization underscores the profound impact these microorganisms have on our health.

The microbiome begins to develop at birth and continues to evolve throughout life, influenced by various factors such as genetics, diet, environment, and lifestyle. Its composition is highly individualized, like a fingerprint, and can vary significantly from person to person. While some bacteria are considered beneficial and promote health, others may be potentially harmful if their balance is disrupted.

The Gut-Brain Connection

The intricate relationship between the gut and the brain, often referred to as the gut-brain axis, highlights the interconnectedness of these two vital systems. The communication occurs through the vagus nerve, a long nerve that extends from the brainstem to the abdomen, allowing bidirectional signaling between the gut and the brain.

Research has revealed that the gut microbiome influences not only digestive processes but also plays a pivotal role in mental health and cognitive function. Imbalances in the microbiome have been linked to conditions such as anxiety, depression, and stress. This intricate connection emphasizes the holistic nature of health, where the well-being of the mind and body are intricately intertwined.

Digestive Harmony: The Role of Gut Microbes in Digestion

At its core, the gut microbiome is a key player in the process of digestion. Microbes aid in breaking down complex carbohydrates, synthesizing certain vitamins, and facilitating nutrient absorption. Moreover, they contribute to the fermentation of dietary fibers, producing short-chain fatty acids (SCFAs), which have been associated with numerous health benefits.

Probiotic bacteria, such as Lactobacillus and Bifidobacterium species, are particularly important for maintaining digestive harmony. These beneficial microbes help prevent the overgrowth of harmful bacteria, support the integrity of the intestinal lining, and contribute to the production of enzymes that aid in digestion.

An imbalance in the gut microbiome, often referred to as dysbiosis, can lead to digestive issues such as bloating, gas, constipation, or diarrhea. Chronic digestive conditions, including irritable bowel syndrome (IBS) and inflammatory bowel diseases (IBD), have been linked to disruptions in the delicate balance of the microbiome.

Immune System Guardians: The Microbiome's Role in Immunity

The gut serves as a frontline defense in our immune system, and the microbiome plays a crucial role in regulating immune function. Beneficial bacteria within the gut help train the immune system to distinguish between

harmful pathogens and harmless substances, contributing to a balanced and responsive immune system.

Additionally, the gut microbiome produces antimicrobial peptides and other compounds that inhibit the growth of harmful bacteria and viruses. A healthy microbiome acts as a fortress, bolstering the body's defenses against infections and promoting resilience to external threats.

Microbes as Metabolic Managers: Influence on Weight and Metabolism

Emerging research has shed light on the role of the gut microbiome in regulating metabolism and weight. The composition of the microbiome has been linked to body weight, and certain microbes may influence energy balance and fat storage. An imbalance in the microbiome has been associated with conditions like obesity and metabolic syndrome.

The fermentation of dietary fibers by gut microbes produces SCFAs, which not only contribute to gut health but also play a role in metabolic processes. SCFAs can influence appetite, regulate insulin sensitivity, and impact the storage of fat in the body. Understanding these intricate connections offers new insights into potential strategies for weight management and metabolic health.

Guardians of Gut Health: Probiotics and Prebiotics

Maintaining a healthy gut involves cultivating a balanced microbiome, and two key players in this endeavor are probiotics and prebiotics.

Live bacteria known as probiotics can benefit one's health if consumed in large enough amounts. Found in fermented foods such as yogurt, kefir, sauerkraut, and kimchi, probiotics contribute to the diversity and balance of the gut microbiome. They can also be taken in supplement form, providing a convenient way to support gut health.

Prebiotics: These are non-digestible fibers that serve as a food source for beneficial bacteria in the gut. Prebiotics are found in foods such as garlic, onions, leeks, bananas, and whole grains. Including these fiber-rich foods in the diet nourishes the microbiome, promoting the growth of beneficial bacteria.

The Impact of Diet on the Microbiome

Diet is a key factor in determining the gut microbiome's composition. A diet rich in diverse, plant-based foods provides an array of nutrients and fibers that support a flourishing microbiome. On the other hand, a diet high in processed foods, sugar, and saturated fats can contribute to an imbalance in the microbiome, fostering the growth of potentially harmful bacteria.

Including a variety of fruits, vegetables, whole grains, legumes, and fermented foods in your diet provides essential nutrients for both you and your microbiome. Diversity in food choices translates to diversity in the microbiome, creating an environment that supports optimal health.

Environmental Influences on the Microbiome

Beyond diet, various environmental factors can impact the microbiome. Antibiotic use, for example, can disrupt the balance of gut bacteria, potentially leading to dysbiosis. Additionally, exposure to environmental toxins and pollutants may influence the composition of the microbiome.

While some factors are beyond our control, maintaining a healthy lifestyle can positively influence the microbiome. Regular exercise, adequate sleep, and stress management contribute to a balanced and resilient gut environment.

The Gut as a Therapeutic Target: Microbiome and Disease

The role of the microbiome in health extends beyond digestion, immunity, and metabolism. Research has unveiled connections between the microbiome and a myriad of diseases, ranging from gastrointestinal disorders to systemic conditions.

Inflammatory bowel diseases (IBD), including Crohn's disease and ulcerative colitis, are characterized by chronic inflammation of the digestive tract. The microbiome is intricately involved in the development and progression of these conditions, and therapeutic approaches targeting the microbiome are being explored as potential avenues for managing IBD.

Similarly, research has linked imbalances in the gut microbiome to conditions such as irritable bowel syndrome (IBS), allergies, and even neurodegenerative diseases like Parkinson's and Alzheimer's. Understanding these connections opens up new possibilities for targeted interventions that focus on restoring and maintaining a healthy microbiome.

Practical Steps for Supporting a Healthy Gut

Diversify Your Diet: Include a variety of fruits, vegetables, whole grains, legumes, and fermented foods in your meals to promote microbiome diversity.

Probiotic-Rich Foods: Incorporate fermented foods like yogurt, kefir, sauerkraut, kimchi, and miso into your diet for a natural source of probiotics.

Prebiotic Foods: Consume foods rich in prebiotic fibers, such as garlic, onions, leeks, bananas, and whole grains, to nourish beneficial bacteria.

Limit Processed Foods: Minimize the intake of processed foods, refined sugars, and saturated fats, as they can negatively impact the balance of the microbiome.

Stay Hydrated: Adequate water intake is essential for maintaining a healthy digestive system and supporting the functions of the gut.

Manage Stress: Practice stress-reducing techniques such as meditation, deep breathing, or yoga, as chronic stress can negatively impact the gut.

Limit Antibiotic Use: Use antibiotics judiciously and only as prescribed by a healthcare professional to minimize disruption to the microbiome.

Regular Exercise: Engage in regular physical activity, as exercise has been shown to positively influence the composition and diversity of the gut microbiome.

Conclusion: Nurturing Your Microbial Garden

As we delve into the intricate world of the gut microbiome, it becomes clear that its health is intricately linked to our overall well-being. From digestion and immune function to mental health and disease prevention, the microbiome plays a pivotal role in shaping our health trajectory.

Embracing a holistic approach to health involves recognizing the symbiotic relationship we share with our microbial counterparts. By nourishing our microbiome through a diverse and nutrient-rich diet, practicing mindful lifestyle habits, and understanding the impact of environmental factors, we can cultivate a thriving microbial garden within.

The journey to a healthy gut is not a one-size-fits-all endeavor. Each person's microbiome is unique, shaped by a myriad of factors, and responding to individualized interventions. As we continue to unravel the mysteries of the gut microbiome, it is a testament to the interconnectedness of our body's systems and the profound impact that conscious choices can have on our health.

So, let this exploration be an invitation—a call to action to nurture and celebrate the symbiotic dance within. As you embark on the path to a healthier gut, remember that every bite, every mindful moment, contributes to the well-being of your microbial allies. The journey is not just about understanding the microbiome; it's about fostering a relationship with it—a partnership that holds the key to unlocking the full potential of your health.

Breakfast

Gut-Friendly Smoothies

Berry Bliss Smoothie

Ingredients:

1 cup mixed berries (blueberries, strawberries, raspberries)

1 banana

1/2 cup Greek yogurt (probiotic-rich)

1 tablespoon chia seeds (prebiotic)

1 cup almond milk

Ice cubes

Instructions:

1. Blend all ingredients until smooth.
2. Pour into a glass and enjoy the burst of antioxidants and probiotics.

Green Goddess Smoothie

Ingredients:

1 cup spinach or kale (rich in fiber)

1/2 cucumber, peeled

1/2 green apple

1/2 avocado

1 tablespoon flaxseeds (prebiotic)

1 cup coconut water

Instructions:

1. Blend until creamy.
2. Pour into a glass, and relish the nourishing green goodness packed with fiber and healthy fats.

Protein-Packed Peanut Butter Smoothie

Ingredients:

2 tablespoons peanut butter

1 banana

1/2 cup plain yogurt (probiotic)

1 scoop protein powder

1 tablespoon honey

1 cup almond milk

Instructions:

1. Blend all ingredients until well combined.
2. Sip on this satisfying smoothie for a protein boost and gut-friendly probiotics.

Tropical Turmeric Smoothie

Ingredients:

1 cup pineapple chunks

1/2 mango

1/2 teaspoon turmeric

1 tablespoon grated ginger

1/2 cup coconut milk

1 tablespoon hemp seeds (prebiotic)

Instructions:

1. Blend until smooth. This tropical delight is not only refreshing but also provides anti-inflammatory benefits.

Chia Seed Pudding Parfait

Ingredients:

2 tablespoons chia seeds

1 cup almond milk

1 teaspoon vanilla extract

1/2 cup mixed berries

1/4 cup granola (optional)

Instructions:

1. Mix chia seeds, almond milk, and vanilla extract.
2. Let it sit in the refrigerator overnight.
3. In the morning, layer with berries and granola.

Probiotic Yogurt Bowl

Ingredients:

1 cup Greek yogurt (probiotic)

1/2 cup mixed berries

1 tablespoon honey

1/4 cup nuts and seeds (almonds, walnuts, chia seeds)

Instructions:

1. Mix Greek yogurt with honey.
2. Top with berries and a sprinkle of nuts and seeds for a protein-packed breakfast.

Oatmeal Banana Pancakes

Ingredients:

1 cup rolled oats

2 ripe bananas

2 eggs

1/2 teaspoon cinnamon

1/2 teaspoon baking powder

1/4 cup almond milk

Instructions:

1. Blend all ingredients until smooth.
2. Cook as you would regular pancakes.
3. Serve with a dollop of Greek yogurt for an extra probiotic kick.

Coconut and Mango Overnight Oats

Ingredients:

1/2 cup rolled oats

1/2 cup coconut milk

1/2 cup diced mango

1 tablespoon shredded coconut (prebiotic)

1 tablespoon chia seeds

Instructions:

1. Mix all ingredients in a jar and refrigerate overnight.
2. In the morning, enjoy a delicious and gut-friendly breakfast on the go.

Turmeric and Ginger Infused Overnight Chia Pudding

Ingredients:

2 tablespoons chia seeds

1 cup coconut milk

1/2 teaspoon ground turmeric

1/2 teaspoon grated ginger

1 tablespoon maple syrup

Sliced kiwi and a sprinkle of pumpkin seeds for topping

Instructions:

1. In a bowl, whisk together chia seeds, coconut milk, turmeric, ginger, and maple syrup.
2. Refrigerate overnight.
3. Top with sliced kiwi and pumpkin seeds before serving for a refreshing and anti-inflammatory breakfast.

Blueberry Almond Butter Smoothie Bowl

Ingredients:

1 cup frozen blueberries

1 tablespoon almond butter

1/2 cup Greek yogurt (probiotic)

1 tablespoon honey

1/4 cup granola

Sliced almonds for topping

Instructions:

1. Blend frozen blueberries, almond butter, Greek yogurt, and honey until smooth.

2. Pour into a bowl and top with granola and sliced almonds for added crunch and nutritional benefits.

Overnight Oats with Probiotic Toppings

Classic Probiotic Parfait

Ingredients:

1/2 cup rolled oats

1/2 cup Greek yogurt (probiotic)

1/2 cup almond milk

1 tablespoon chia seeds (prebiotic)

1 teaspoon honey

Fresh berries for topping

Instructions:

1. Mix rolled oats, Greek yogurt, almond milk, chia seeds, and honey in a jar.
2. Refrigerate overnight.
3. Top with fresh berries before serving for a classic and probiotic-rich parfait.

Mango Coconut Kefir Overnight Oats

Ingredients:

1/2 cup rolled oats

1/2 cup coconut kefir (probiotic)

1/2 cup diced mango

1 tablespoon shredded coconut (prebiotic)

1 tablespoon sliced almonds

Drizzle of agave syrup

Instructions:

1. Combine rolled oats, coconut kefir, diced mango, shredded coconut, and sliced almonds in a jar.
2. Refrigerate overnight.
3. Drizzle with agave syrup before serving for a tropical delight.

Vanilla Yogurt and Berry Overnight Oat

Ingredients:

1/2 cup rolled oats

1/2 cup vanilla yogurt (probiotic)

1/2 cup mixed berries

1 tablespoon flaxseeds (prebiotic)

1 tablespoon maple syrup

Granola for topping

Instructions:

1. Mix rolled oats, vanilla yogurt, mixed berries, flaxseeds, and maple syrup in a jar.
2. Refrigerate overnight. Top with granola before serving for added crunch.

Peach and Probiotic Yogurt Overnight Oats

Ingredients:

1/2 cup rolled oats

1/2 cup peach-flavored yogurt (probiotic)

1/2 cup diced fresh peaches

1 tablespoon chia seeds (prebiotic)

1 tablespoon chopped pecans

Drizzle of honey

Instructions:

1. Combine rolled oats, peach-flavored yogurt, diced peaches, chia seeds, and chopped pecans in a jar. Refrigerate overnight.

2. Drizzle with honey before serving for a sweet and probiotic-rich treat.

Blueberry Probiotic Overnight Oats

Ingredients:

1/2 cup rolled oats

1/2 cup blueberry yogurt (probiotic)

1/2 cup blueberries

1 tablespoon hemp seeds (prebiotic)

1 tablespoon almond butter

Sliced bananas for topping

Instructions:

1. Mix rolled oats, blueberry yogurt, blueberries, hemp seeds, and almond butter in a jar.
2. Refrigerate overnight.
3. Top with sliced bananas before serving for a delightful blueberry twist.

Chocolate Banana Kefir Overnight Oats

Ingredients:

1/2 cup rolled oats

1/2 cup chocolate-flavored kefir (probiotic)

1 ripe banana, mashed

1 tablespoon cocoa powder

1 tablespoon chopped walnuts (prebiotic)

A sprinkle of dark chocolate chips

Instructions:

1. Combine rolled oats, chocolate-flavored kefir, mashed banana, cocoa powder, and chopped walnuts in a jar.
2. Refrigerate overnight.
3. Sprinkle with dark chocolate chips before serving for a decadent yet gut-friendly option.

High-Fiber Breakfast Bowls

Triple Berry Chia Seed Pudding Bowl

Ingredients:

1/2 cup chia seeds

1.5 cups almond milk

1 cup mixed berries (strawberries, blueberries, raspberries)

1 tablespoon honey

1/4 cup granola (fiber-rich)

Sliced almonds for topping

Instructions:

1. Mix chia seeds and almond milk
2. Let it sit in the refrigerator for a few hours or overnight.
3. In the morning, layer chia pudding with mixed berries, honey, granola, and sliced almonds for a fiber-packed delight.

Apple Cinnamon Quinoa Breakfast Bowl

Ingredients:

1/2 cup cooked quinoa

1 apple, diced

1 tablespoon almond butter

1/2 teaspoon cinnamon

1 tablespoon chopped walnuts (fiber-rich)

Greek yogurt for topping

Instructions:

1. Mix cooked quinoa with diced apple, almond butter, cinnamon, and chopped walnuts.
2. Top with a dollop of Greek yogurt for added creaminess and protein.

Mango Coconut Oatmeal Bowl

Ingredients:

1/2 cup rolled oats

1/2 cup coconut milk

1/2 mango, diced

1 tablespoon shredded coconut (fiber-rich)

1 tablespoon chia seeds

A drizzle of agave syrup

Instructions:

1. Cook rolled oats with coconut milk.
2. Top with diced mango, shredded coconut, chia seeds, and a drizzle of agave syrup for a tropical and fiber-packed breakfast.

Spinach and Mushroom Breakfast Quinoa Bowl

Ingredients:

1/2 cup cooked quinoa

Handful of fresh spinach

1/2 cup sautéed mushrooms

1 poached egg

Salt and pepper to taste

Sprinkle of feta cheese (optional)

Instructions:

1. Mix cooked quinoa with fresh spinach and sautéed mushrooms.
2. Top with a poached egg, salt, pepper, and a sprinkle of feta cheese for a savory, fiber-rich breakfast.

Banana Walnut Overnight Oats Bowl

Ingredients:

1/2 cup rolled oats

1/2 cup almond milk

1 banana, sliced

1 tablespoon chopped walnuts (fiber-rich)

1 tablespoon flaxseeds

A dash of cinnamon

Instructions:

1. Mix rolled oats with almond milk, sliced banana, chopped walnuts, flaxseeds, and cinnamon.
2. Refrigerate overnight and enjoy a creamy and fiber-packed breakfast in the morning.

Berry Almond Chia Bowl

Ingredients:

1/2 cup rolled oats

1 cup almond milk

2 tablespoons chia seeds

1/2 cup mixed berries (blueberries, strawberries, raspberries)

1 tablespoon almond butter

1 tablespoon sliced almonds

Drizzle of honey

Instructions:

Combine rolled oats, almond milk, and chia seeds in a bowl.

Let it sit in the refrigerator overnight.

In the morning, top with mixed berries, almond butter, sliced almonds, and a drizzle of honey.

Tropical Quinoa Breakfast Bowl

Ingredients:

1/2 cup cooked quinoa

1/2 cup coconut milk

1/2 cup diced mango

1/4 cup pineapple chunks

1 tablespoon shredded coconut

1 tablespoon chopped macadamia nuts

A sprinkle of chia seeds

Instructions:

1. Mix cooked quinoa with coconut milk.
2. Top with diced mango, pineapple chunks, shredded coconut, chopped macadamia nuts, and chia seeds.

Apple Cinnamon Oat Bran Bowl

Ingredients:

1/2 cup oat bran

1 cup water or milk

1/2 apple, diced

1 tablespoon raisins

1/2 teaspoon cinnamon

1 tablespoon chopped walnuts

Drizzle of maple syrup

Instructions:

1. Cook oat bran in water or milk until creamy.
2. Top with diced apples, raisins, cinnamon, chopped walnuts, and a drizzle of maple syrup.

Spinach and Feta Breakfast Bulgur Bowl

Ingredients:

1/2 cup cooked bulgur

Handful of fresh spinach

1/4 cup crumbled feta cheese

1/2 cherry tomatoes, halved

1 tablespoon pumpkin seeds

Olive oil and balsamic glaze for drizzling

Instructions:

1. Mix cooked bulgur with fresh spinach.
2. Top with crumbled feta, cherry tomatoes, pumpkin seeds, and drizzle with olive oil and balsamic glaze.

Banana Walnut Buckwheat Bowl

Ingredients:

1/2 cup cooked buckwheat groats

1/2 banana, sliced

1 tablespoon chopped walnuts

1 tablespoon hemp seeds

1 tablespoon Greek yogurt

Drizzle of honey

Instructions:

1. Combine cooked buckwheat with banana slices, chopped walnuts, hemp seeds, and a dollop of Greek yogurt.
2. Drizzle with honey before serving.

Protein-Packed Black Bean and Avocado Bowl

Ingredients:

1/2 cup black beans (canned or cooked)

1/2 cup quinoa, cooked

1/4 avocado, sliced

1/2 cup cherry tomatoes, halved

1 tablespoon cilantro, chopped

Squeeze of lime juice

Salt and pepper to taste

Instructions:

1. Mix black beans and cooked quinoa.
2. Top with avocado slices, cherry tomatoes, chopped cilantro, and a squeeze of lime juice.
3. Season with salt and pepper.

Lunches

Gut-Healing Soups and Stews
Turmeric and Ginger Detox Soup

Ingredients:

1 tablespoon olive oil

1 onion, chopped

3 cloves garlic, minced

1 tablespoon fresh ginger, grated

1 tablespoon fresh turmeric, grated

4 cups vegetable broth

2 carrots, sliced

1 cup kale, chopped

Salt and pepper to taste

Instructions:

1. In a large pot, sauté onions, garlic, ginger, and turmeric in olive oil until fragrant.
2. Add vegetable broth, carrots, and kale. Simmer until vegetables are tender.
3. Season with salt and pepper to taste.

Probiotic-rich Miso Soup

Ingredients:

4 cups vegetable broth

2 tablespoons miso paste

1 cup tofu, cubed

1 cup seaweed, sliced

2 green onions, chopped

Instructions:

Heat vegetable broth in a pot.

Dissolve miso paste in a bit of broth before adding it to the pot.

Add tofu and seaweed, simmer until heated through.

Garnish with green onions before serving.

Bone Broth Vegetable Stew

Ingredients:

4 cups bone broth

1 cup sweet potatoes, diced

1 cup zucchini, sliced

1 cup broccoli florets

1 cup spinach

1 teaspoon thyme

Salt and pepper to taste

Instructions:

1. Bring bone broth to a simmer in a pot.
2. Add sweet potatoes, zucchini, and broccoli. Cook until vegetables are tender.
3. Stir in spinach, thyme, salt, and pepper before serving.

Cabbage and Turmeric Healing Soup:

Ingredients:

1 tablespoon coconut oil

1 onion, diced

3 cups cabbage, shredded

1 tablespoon turmeric powder

4 cups vegetable broth

1 cup celery, chopped

1 cup carrots, sliced

Instructions:

1. Sauté onions in coconut oil until softened.
2. Add cabbage, turmeric, vegetable broth, celery, and carrots. Simmer until vegetables are tender.

Lentil and Kale Gut Balancing Stew

Ingredients:

1 cup lentils, rinsed

1 onion, chopped

3 cloves garlic, minced

1 cup kale, chopped

4 cups vegetable broth

1 teaspoon cumin

Salt and pepper to taste

Instructions:

1. In a pot, sauté onions and garlic until translucent.
2. Add lentils, vegetable broth, kale, cumin, salt, and pepper. Simmer until lentils are cooked.

Carrot and Ginger Immune-Boosting Soup

Ingredients:

1 tablespoon olive oil

1 onion, chopped

4 cups carrots, sliced

1 tablespoon fresh ginger, grated

4 cups vegetable broth

1 cup coconut milk

Salt and pepper to taste

Instructions:

1. Sauté onions in olive oil until soft.
2. Add carrots, ginger, vegetable broth, and simmer until carrots are tender.
3. Blend the soup until smooth, stir in coconut milk, and season with salt and pepper.

Quinoa and Vegetable Gut Repair Stew:

Ingredients:

1 cup quinoa, rinsed

1 onion, diced

3 cloves garlic, minced

1 cup bell peppers, chopped

4 cups vegetable broth

1 cup green beans, chopped

1 teaspoon turmeric

Salt and pepper to taste

Instructions:

1. Sauté onions and garlic until softened.
2. Add quinoa, bell peppers, vegetable broth, green beans, turmeric, salt, and pepper. Simmer until quinoa is cooked.

Chickpea and Spinach Gut Rejuvenation Stew

Ingredients:

1 tablespoon olive oil

1 onion, finely chopped

3 cloves garlic, minced

1 can (15 oz) chickpeas, drained and rinsed

1 cup carrots, diced

1 cup celery, chopped

4 cups vegetable broth

1 teaspoon cumin

1 teaspoon coriander

2 cups spinach, chopped

Salt and pepper to taste

Instructions:

1. Sauté onions and garlic in olive oil until softened.
2. Add chickpeas, carrots, celery, vegetable broth, cumin, and coriander. Simmer until vegetables are tender.
3. Stir in chopped spinach and season with salt and pepper.

Sweet Potato and Coconut Gut Soothing Soup

Ingredients:

2 tablespoons coconut oil

1 onion, diced

3 cups sweet potatoes, peeled and diced

1 tablespoon fresh ginger, grated

1 can (14 oz) coconut milk

4 cups vegetable broth

1 teaspoon turmeric

1 teaspoon cinnamon

Salt and pepper to taste

Instructions:

1. In a pot, sauté onions in coconut oil until translucent.
2. Add sweet potatoes, ginger, coconut milk, vegetable broth, turmeric, and cinnamon. Simmer until sweet potatoes are tender.
3. Season with salt and pepper before serving.

Protein-Packed Quinoa and Kale Soup

Ingredients:

1 tablespoon olive oil

1 onion, chopped

3 cloves garlic, minced

1 cup quinoa, rinsed

4 cups vegetable broth

2 cups kale, chopped

1 can (15 oz) cannellini beans, drained and rinsed

1 teaspoon dried thyme

Salt and pepper to taste

Instructions:

1. Sauté onions and garlic in olive oil until softened.
2. Add quinoa, vegetable broth, kale, cannellini beans, thyme, salt, and pepper. Simmer until quinoa is cooked and kale is tender

Fiber-Rich Salads with Probiotic Dressings

Quinoa and Kale Power Salad with Yogurt-Tahini Dressing

Salad:

2 cups cooked quinoa

2 cups kale, finely chopped

1 cucumber, diced

1 cup cherry tomatoes, halved

Dressing:

1/2 cup Greek yogurt

2 tablespoons tahini

1 tablespoon lemon juice

1 clove garlic, minced

Salt and pepper to taste

Broccoli and Chickpea Crunch Salad with Kimchi Dressing

Salad:

3 cups broccoli florets, blanched

1 can (15 oz) chickpeas, drained and rinsed

1 cup red cabbage, shredded

1/2 cup almonds, chopped

Dressing:

1/4 cup kimchi, chopped

2 tablespoons olive oil

1 tablespoon rice vinegar

1 teaspoon honey

Salt and pepper to taste

Mixed Bean and Avocado Salad with Miso-Ginger Dressing

Salad:

1 cup mixed beans (kidney, black, chickpeas), cooked

1 avocado, diced

1 cup cherry tomatoes, halved

1/2 red onion, finely chopped

Dressing:

2 tablespoons miso paste

1 tablespoon grated ginger

2 tablespoons rice vinegar

1 tablespoon soy sauce

1 tablespoon sesame oil

Spinach and Berry Bliss Salad with Kefir-Citrus Dressing

Salad:

3 cups fresh spinach leaves

1 cup strawberries, sliced

1/2 cup blueberries

1/4 cup almonds, sliced

Dressing:

1/2 cup kefir

2 tablespoons orange juice

Zest of one orange

1 tablespoon honey

Salt to taste

Sauerkraut and Apple Slaw with Mustard-Maple Vinaigrette

Salad:

2 cups cabbage, shredded

1 cup sauerkraut

1 apple, julienned

1/4 cup walnuts, chopped

Dressing:

2 tablespoons Dijon mustard

1 tablespoon apple cider vinegar

1 tablespoon maple syrup

2 tablespoons olive oil

Salt and pepper to taste

Fermented Beet and Lentil Salad with Yogurt-Dill Dressing

Salad:

2 cups cooked lentils

1 cup fermented beets, chopped

1 cucumber, diced

1/4 cup feta cheese, crumbled

Dressing:

1/2 cup Greek yogurt

1 tablespoon fresh dill, chopped

1 tablespoon lemon juice

Salt and pepper to taste

Asian-Inspired Seaweed and Edamame Salad with Miso-Sesame Dressing

Salad:

1 cup dried seaweed, rehydrated

1 cup edamame, steamed

1 carrot, julienned

1/4 cup green onions, sliced

Dressing:

2 tablespoons miso paste

1 tablespoon sesame oil

1 tablespoon rice vinegar

1 teaspoon soy sauce

1 teaspoon honey

Grilled Veggie and Quinoa Salad with Yogurt-Tzatziki Dressing

Salad:

2 cups cooked quinoa

1 zucchini, sliced

1 red bell pepper, sliced

1 eggplant, diced

1 cup cherry tomatoes, halved

Dressing:

1/2 cup Greek yogurt

2 tablespoons cucumber, finely chopped

1 tablespoon fresh dill, chopped

1 tablespoon lemon juice

Salt and pepper to taste

Mexican Black Bean and Corn Salad with Fermented Jalapeño Dressing

Salad:

2 cups black beans, cooked

1 cup corn kernels, grilled

1 red onion, finely chopped

1 cup cherry tomatoes, halved

Dressing:

2 tablespoons fermented jalapeños, chopped

2 tablespoons lime juice

2 tablespoons olive oil

1 teaspoon cumin

Salt and pepper to taste

These salads bring a diverse array of flavors, textures, and nutrients to the table, enhancing your gut health with fiber-rich ingredients and probiotic-packed dressings. Enjoy the delightful combinations as you nourish your body and support your digestive well-being.

Fermented Sides for Gut Health
Kimchi

Ingredients:

1 Napa cabbage, shredded

1 daikon radish, julienned

3 cloves garlic, minced

1 tablespoon ginger, grated

2 tablespoons Korean red pepper flakes

2 tablespoons fish sauce

Salt

Instructions:

1. Combine cabbage, radish, garlic, ginger, red pepper flakes, and fish sauce in a bowl.
2. Massage the ingredients with salt.
3. Pack the mixture tightly into a jar and let it ferment for about a week.

Sauerkraut

Ingredients:

1 head green cabbage, shredded

1 tablespoon caraway seeds

1 tablespoon sea salt

Instructions:

1. Combine cabbage, caraway seeds, and sea salt in a bowl.
2. Massage the mixture until the cabbage releases its juices.
3. Pack the cabbage tightly into a jar, ensuring it's submerged in its own juices.
4. Allow it to ferment for 2-3 weeks.

Fermented Pickles

Ingredients:

1 pound pickling cucumbers, sliced

1 tablespoon dill seeds

2 cloves garlic, sliced

1 tablespoon sea salt

Water

Instructions:

1. Pack cucumbers, dill seeds, and garlic into a jar.
2. Dissolve sea salt in water and pour over the cucumbers until submerged.
3. Allow it to ferment for about a week.

Fermented Carrots with Ginger

Ingredients:

1 pound carrots, peeled and sliced

1 tablespoon fresh ginger, grated

1 tablespoon sea salt

Instructions:

1. Mix carrots, ginger, and sea salt in a bowl.

2. Pack the mixture into a jar and press it down to release juices.
3. Let it ferment for about 2 weeks.

Pickled Red Onions

Ingredients:

2 red onions, thinly sliced

1 cup apple cider vinegar

1 tablespoon sugar

1 teaspoon sea salt

Instructions:

1. Dissolve sugar and sea salt in apple cider vinegar.
2. Pour the mixture over the red onions in a jar.
3. Allow it to pickle for at least 24 hours.

Fermented Beet Kvass:

Ingredients:

2 large beets, peeled and chopped

1 tablespoon sea salt

Water

Instructions:

1. Place beets and sea salt in a jar.
2. Fill the jar with water and cover.
3. Allow it to ferment for about a week.

Lacto-Fermented Jalapeño Slices

Ingredients:

1 cup jalapeño peppers, sliced

2 cloves garlic, sliced

1 tablespoon sea salt

Water

Instructions:

1. Pack jalapeños and garlic into a jar.
2. Dissolve sea salt in water and pour over the peppers.
3. Let it ferment for about a week.

Fermented Radishes with Herbs

Ingredients:

1 bunch radishes, sliced

1 tablespoon fresh dill, chopped

1 tablespoon sea salt

Instructions:

1. Mix radishes, dill, and sea salt in a bowl.
2. Pack the mixture into a jar and press it down.
3. Allow it to ferment for about 2 weeks.

Cultured Green Beans

Ingredients:

1 pound green beans, trimmed

2 cloves garlic, sliced

1 tablespoon sea salt

Water

Instructions:

1. Pack green beans and garlic into a jar.
2. Dissolve sea salt in water and pour over the beans.
3. Let it ferment for about a week.

Fermented Cilantro-Lime Slaw

Ingredients:

1/2 head green cabbage, shredded

1/2 cup cilantro, chopped

Juice of 2 limes

1 tablespoon sea salt

Instructions:

1. Combine cabbage, cilantro, lime juice, and sea salt in a bowl.
2. Massage the mixture until it releases its juices.
3. Pack it tightly into a jar and let it ferment for about 2 weeks.

These fermented sides not only add a burst of flavor to your meals but also contribute to a healthy gut by providing beneficial probiotics. Enjoy incorporating these delicious and nutritious fermented foods into your diet!

Dinners

Lean Proteins with Probiotic Marinades

Grilled Lemon-Herb Salmon with Yogurt-Dill Marinade

Ingredients:

4 salmon fillets

1/2 cup Greek yogurt

2 tablespoons fresh dill, chopped

Zest and juice of 1 lemon

2 cloves garlic, minced

Salt and pepper to taste

Instructions:

1. In a bowl, mix yogurt, dill, lemon zest, lemon juice, garlic, salt, and pepper.
2. Marinate salmon fillets in the mixture for at least 30 minutes.
3. Grill the salmon until cooked to your liking.

Turmeric-Ginger Chicken Skewers with Fermented Mango Salsa

Ingredients:

1 pound chicken breast, cubed

1 tablespoon turmeric powder

1 tablespoon fresh ginger, grated

2 tablespoons olive oil

Salt and pepper to taste

For the Mango Salsa:

1 ripe mango, diced

1/4 cup red onion, finely chopped

2 tablespoons cilantro, chopped

2 tablespoons fermented jalapeños, chopped

Instructions:

1. In a bowl, combine chicken, turmeric, ginger, olive oil, salt, and pepper. Marinate for at least 1 hour.
2. After marinating, thread the chicken onto skewers and cook over a grill.
3. Mix mango, red onion, cilantro, and jalapeños for the salsa. Serve over chicken skewers.

Balsamic-Kefir Marinated Turkey Cutlets

Ingredients

4 turkey cutlets

1/2 cup balsamic vinegar

1/4 cup kefir

2 tablespoons honey

2 cloves garlic, minced

Salt and pepper to taste

Instructions:

1. In a bowl, whisk together balsamic vinegar, kefir, honey, garlic, salt, and pepper.
2. Marinate turkey cutlets in the mixture for at least 1 hour.
3. Grill or pan-sear the turkey cutlets until fully cooked.

Miso-Soy Marinated Tofu Stir-Fry:

Ingredients:

1 block extra-firm tofu, pressed and cubed

2 tablespoons miso paste

2 tablespoons soy sauce

1 tablespoon sesame oil

1 tablespoon rice vinegar

1 tablespoon maple syrup

2 cloves garlic, minced

1 tablespoon ginger, grated

Vegetables of your choice (broccoli, bell peppers, snap peas)

Brown rice or quinoa for serving

Instructions:

1. In a bowl, whisk together miso paste, soy sauce, sesame oil, rice vinegar, maple syrup, garlic, and ginger.
2. Marinate tofu cubes in the mixture for at least 30 minutes.
3. Stir-fry marinated tofu and vegetables until cooked. Serve over brown rice or quinoa.

Probiotic Yogurt Marinated Shrimp Skewers

Ingredients:

1 pound large shrimp, peeled and deveined

1 cup plain yogurt

2 tablespoons lemon juice

1 tablespoon smoked paprika

1 teaspoon cumin

2 cloves garlic, minced

Salt and pepper to taste

Instructions:

1. In a bowl, mix yogurt, lemon juice, smoked paprika, cumin, garlic, salt, and pepper.
2. Shrimp should be marinated in the marinade for at least half an hour.
3. Thread marinated shrimp onto skewers and grill until opaque and cooked through.

Yogurt and Herb Marinated Grilled Chicken Breast

Ingredients

4 boneless, skinless chicken breasts

1 cup plain yogurt

2 tablespoons fresh basil, chopped

2 tablespoons fresh parsley, chopped

1 tablespoon lemon zest

2 cloves garlic, minced

Salt and pepper to taste

Instructions:

1. In a bowl, combine yogurt, basil, parsley, lemon zest, garlic, salt, and pepper.
2. Marinate chicken breasts in the mixture for at least 1 hour.
3. Grill the chicken until fully cooked.

Kombucha-Glazed Teriyaki Salmon

Ingredients:

4 salmon fillets

1/2 cup kombucha

1/4 cup soy sauce

2 tablespoons honey

1 tablespoon ginger, grated

1 clove garlic, minced

2 green onions, chopped (for garnish)

Instructions:

1. In a bowl, whisk together kombucha, soy sauce, honey, ginger, and garlic.

2. Marinate salmon fillets in the mixture for at least 30 minutes.
3. Bake or grill the salmon until it flakes easily. Garnish with chopped green onions.

Lemony Yogurt-Marinated Swordfish Steaks

Ingredients:

4 swordfish steaks

1 cup Greek yogurt

Juice of 2 lemons

1 tablespoon fresh thyme, chopped

1 teaspoon smoked paprika

Salt and pepper to taste

Instructions:

1. In a bowl, mix yogurt, lemon juice, thyme, smoked paprika, salt, and pepper.
2. Marinate swordfish steaks in the mixture for at least 1 hour.
3. Grill or pan-sear the swordfish until cooked to your liking.

Probiotic Buttermilk Marinated Turkey Burgers

Ingredients:

1 pound ground turkey

1 cup buttermilk

2 tablespoons Dijon mustard

1 tablespoon Worcestershire sauce

1 teaspoon smoked paprika

2 cloves garlic, minced

Salt and pepper to taste

Instructions:

1. In a bowl, combine ground turkey, buttermilk, Dijon mustard, Worcestershire sauce, smoked paprika, garlic, salt, and pepper.
2. Form the mixture into burger patties and let them marinate for at least 30 minutes.
3. Grill the turkey burgers until fully cooked.

Lemon-Honey Yogurt Marinated Tofu Steaks
Ingredients:

1 block extra-firm tofu, pressed and sliced into steaks

1/2 cup plain yogurt

Zest and juice of 1 lemon

2 tablespoons honey

1 tablespoon olive oil

1 teaspoon dried oregano

Salt and pepper to taste

Instructions:

1. In a bowl, whisk together yogurt, lemon zest, lemon juice, honey, olive oil, oregano, salt, and pepper.
2. Marinate tofu steaks in the mixture for at least 1 hour.
3. Pan-fry or grill the tofu steaks until golden brown. Serve with your favorite sides.

Gut-Healing Grains and Vegetables

Quinoa and Roasted Vegetable Buddha Bowl

Ingredients:

1 cup cooked quinoa

Assorted roasted vegetables (e.g., sweet potatoes, bell peppers, zucchini)

1/2 cup chickpeas, roasted

Fresh spinach or kale

Tahini dressing

Instructions:

1. Arrange cooked quinoa, roasted vegetables, chickpeas, and fresh greens in a bowl.
2. Drizzle with tahini dressing for a nourishing and gut-healing Buddha bowl.

Brown Rice Stir-Fry with Fermented Tofu

Ingredients:

1 cup cooked brown rice

Mixed stir-fry vegetables (broccoli, carrots, snap peas)

Fermented tofu cubes

Soy sauce or tamari

Sesame oil

Green onions for garnish

Instructions:

1. Stir-fry the mixed vegetables and fermented tofu in sesame oil.
2. Add cooked brown rice and soy sauce.

3. Garnish with green onions for a flavorful and gut-friendly dish.

Barley and Vegetable Soup

Ingredients:

1/2 cup pearled barley

Mixed vegetables (carrots, celery, onions, kale)

Vegetable broth

Garlic and thyme for seasoning

Olive oil

Lemon juice

Instructions:

1. Sauté garlic in olive oil, add vegetables and cook until softened.
2. Add barley, thyme, and vegetable broth. Simmer until barley is tender.
3. Finish with a squeeze of lemon juice for a comforting and gut-healing soup.

Spaghetti Squash Primavera

Ingredients

1 spaghetti squash, roasted and shredded

Mixed vegetables (cherry tomatoes, bell peppers, asparagus)

Garlic and basil for seasoning

Olive oil

Grated Parmesan cheese (optional)

Instructions:

1. Sauté mixed vegetables in olive oil until tender.
2. Add roasted spaghetti squash, garlic, and basil. Toss until well combined.
3. Serve with a sprinkle of grated Parmesan cheese if desired.

Millet and Roasted Root Vegetables Salad

Ingredients:

1 cup cooked millet

Roasted root vegetables (beets, carrots, parsnips)

Arugula or mixed greens

Toasted pumpkin seeds

Balsamic vinaigrette

Instructions:

1. Combine cooked millet, roasted root vegetables, arugula, and toasted pumpkin seeds in a bowl.
2. Drizzle with balsamic vinaigrette for a hearty and gut-healing salad.

Wild Rice and Brussels Sprouts Bowl

Ingredients:

1 cup cooked wild rice

Roasted Brussels sprouts

Pomegranate seeds

Sliced almonds

Lemon-tahini dressing

Instructions:

1. Mix cooked wild rice, roasted Brussels sprouts, pomegranate seeds, and sliced almonds in a bowl.
2. Drizzle with lemon-tahini dressing for a nutrient-packed and gut-healing grain bowl.

Farro and Mediterranean Vegetable Skewers

Ingredients:

1 cup cooked farro

Cherry tomatoes

Zucchini and yellow squash slices

Red onion chunks

Olive oil, garlic, and oregano for seasoning

Lemon wedges for serving

Instructions:

1. Thread cherry tomatoes, zucchini, yellow squash, and red onion onto skewers.
2. Brush with olive oil, minced garlic, and oregano. Grill or bake until vegetables are tender.
3. Serve over cooked farro with a squeeze of lemon.

One-Pot Meals for Easy Digestion

Chicken and Rice Soup

Ingredients:

1 lb chicken thighs, boneless and skinless

1 cup rice

1 onion, chopped

2 carrots, sliced

2 celery stalks, chopped

4 cups chicken broth

1 teaspoon ginger, grated

Salt and pepper to taste

Instructions:

1. In a large pot, combine all ingredients.
2. Bring to a boil, then reduce heat and simmer until chicken is cooked and rice is tender.

Vegetarian Lentil Stew

Ingredients:

1 cup lentils, rinsed

1 onion, diced

2 carrots, diced

2 potatoes, diced

3 cloves garlic, minced

4 cups vegetable broth

1 teaspoon cumin

1 teaspoon coriander

Salt and pepper to taste

Instructions:

1. In a pot, combine all ingredients.
2. Bring to a boil, then simmer until lentils and vegetables are tender.

Salmon and Quinoa Skillet

Ingredients:

1 lb salmon fillets

1 cup quinoa, rinsed

1 zucchini, sliced

1 bell pepper, diced

1 lemon, sliced

2 tablespoons olive oil

2 cloves garlic, minced

Salt and pepper to taste

Instructions:

1. In a skillet, combine quinoa, vegetables, and olive oil.
2. Place salmon fillets on top, add garlic, lemon slices, salt, and pepper.
3. Cover and cook until quinoa is done and salmon is cooked through.

Shrimp and Vegetable Stir-Fry

Ingredients:

1 lb shrimp, peeled and deveined

2 cups broccoli florets

1 bell pepper, sliced

1 cup snap peas

3 tablespoons soy sauce

2 tablespoons honey

1 tablespoon sesame oil

2 cloves garlic, minced

Instructions:

1. Shrimp and vegetables should be stir-fried in a wok or big pan.
2. Add soy sauce, honey, sesame oil, and garlic. Simmer until the veggies are soft and the prawns are pink.

Turkey and Sweet Potato Chili

Ingredients:

1 lb ground turkey

2 sweet potatoes, diced

One can (15 oz.) black beans, drained and rinsed

1 can (15 oz.) diced tomatoes

1 onion, chopped

2 cloves garlic, minced

2 teaspoons chili powder

Salt and pepper to taste

Instructions:

1. In a pot, brown ground turkey with onions and garlic.

2. Add sweet potatoes, black beans, diced tomatoes, chili powder, salt, and pepper. Simmer until sweet potatoes are tender.

Quinoa and Vegetable Paella

Ingredients:

1 cup quinoa, rinsed

1 lb mixed vegetables (bell peppers, peas, artichokes)

1 onion, diced

3 cloves garlic, minced

2 cups vegetable broth

1 teaspoon smoked paprika

1/2 teaspoon saffron threads (optional)

Salt and pepper to taste

Instructions:

1. In a paella pan or large skillet, sauté onions and garlic.
2. Add quinoa, mixed vegetables, vegetable broth, smoked paprika, saffron, salt, and pepper. Simmer until quinoa is cooked.

Coconut Curry Chicken and Rice

Ingredients:

1 lb chicken thighs, boneless and skinless

1 cup basmati rice, rinsed

1 can (14 oz) coconut milk

1 bell pepper, sliced

1 cup snow peas

2 tablespoons curry powder

1 tablespoon fish sauce

Salt and pepper to taste

Instructions:

1. In a pot, brown chicken thighs.
2. Add rice, coconut milk, bell pepper, snow peas, curry powder, fish sauce, salt, and pepper. Simmer until rice is cooked.

Lemon-Garlic Herb Roasted Chicken with Vegetables

Ingredients:

1 whole chicken, cut into pieces

1 lb baby potatoes, halved

1 cup baby carrots

1 lemon, sliced

4 cloves garlic, minced

2 tablespoons olive oil

1 tablespoon fresh rosemary, chopped

Salt and pepper to taste

Instructions:

1. Preheat the oven. In a large baking dish, combine chicken, potatoes, carrots, lemon slices, garlic, olive oil, rosemary, salt, and pepper.
2. Roast in the oven until the chicken is golden and vegetables are tender.

These one-pot meals are not only easy to prepare but also gentle on digestion, making them perfect for a nourishing and satisfying dinner. Enjoy the simplicity and delicious flavors of these recipes!

Snacks

Probiotic Snack Ideas

Greek Yogurt Parfait

Ingredients

1 cup Greek yogurt

1/2 cup granola

1/2 cup mixed berries (blueberries, strawberries, raspberries)

1 tablespoon honey

Instructions:

1. Arrange Greek yogurt, granola, and mixed berries in a glass or bowl.
2. Drizzle honey on top for sweetness. The live cultures in Greek yogurt contribute to the probiotic content.

Kimchi Avocado Toast

Ingredients:

2 slices whole-grain bread, toasted

1 ripe avocado, mashed

1/4 cup kimchi

Sesame seeds for garnish

Instructions:

1. Spread mashed avocado on toasted bread slices.
2. Top with kimchi and sprinkle sesame seeds. Probiotic-rich fermented food is kimchi.

Fermented Pickle Roll-Ups

Ingredients:

4 large lettuce leaves

8 slices nitrate-free turkey or chicken

4 tablespoons cream cheese

1/2 cup fermented pickles, sliced

Instructions:

1. Lay out lettuce leaves and place turkey or chicken slices on each.
2. Spread cream cheese on each slice and add fermented pickle slices.
3. Roll up and secure with toothpicks if needed.

Kefir Smoothie Bowl

Ingredients:

1 cup kefir

1/2 banana, frozen

1/2 cup mixed berries

1 tablespoon chia seeds

Granola for topping

Instructions:

1. Blend kefir, frozen banana, and mixed berries until smooth.
2. Pour into a bowl and top with chia seeds and granola for added crunch and probiotics.

Sauerkraut and Hummus Stuffed Cucumber Boats

Ingredients:

2 large cucumbers, halved and seeded

1/2 cup hummus

1/2 cup sauerkraut

Fresh dill for garnish

Instructions:

1. Scoop out the seeds from cucumber halves.
2. Fill each cucumber boat with hummus and top with sauerkraut. Garnish with fresh dill.

Probiotic Cheese and Crackers

Ingredients:

1 ounce probiotic-rich cheese (such as aged cheddar or gouda)

Whole-grain crackers

Sliced apple or pear

Instructions:

1. Arrange slices of probiotic-rich cheese on whole-grain crackers.
2. Pair with sliced apple or pear for a balanced and tasty snack.

Miso Roasted Almonds

Ingredients:

1 cup raw almonds

1 tablespoon white miso paste

1 tablespoon olive oil

1 teaspoon soy sauce

Instructions:

1. Preheat oven to 350°F (175°C).
2. In a bowl, mix miso paste, olive oil, and soy sauce.
3. Toss raw almonds in the miso mixture, then spread them on a baking sheet.
4. Roast for 12-15 minutes, stirring halfway. Allow to cool before enjoying.

Nut and Seed Mixes for Gut Health

Classic Almond and Pumpkin Seed Mix

Ingredients:

1 cup almonds

1/2 cup pumpkin seeds

1/4 teaspoon sea salt

Instructions:

1. In a dry pan, lightly toast almonds and pumpkin seeds over medium heat until golden.

2. Sprinkle with sea salt and let it cool before storing.

Walnut and Chia Seed Crunch

Ingredients:

1 cup walnuts

2 tablespoons chia seeds

1 tablespoon maple syrup

1/4 teaspoon cinnamon

Instructions:

1. Toss walnuts and chia seeds in a bowl.
2. Drizzle with maple syrup, sprinkle with cinnamon, and mix well.
3. Spread on a baking sheet and bake at 325°F (163°C) for 10-12 minutes. Let it cool before serving.

Antioxidant Berry Bliss Mix

Ingredients:

1/2 cup almonds

1/2 cup cashews

1/4 cup dried blueberries

1/4 cup dried cranberries

1/4 cup goji berries

Instructions:

1. Combine almonds, cashews, and dried berries in a bowl.
2. Mix well and portion into snack-sized containers.

Probiotic Pistachio and Dark Chocolate Delight

Ingredients:

1 cup pistachios

1/4 cup dark chocolate chips

1/4 cup dried apricots, chopped

1/4 teaspoon sea salt

Instructions:

1. In a bowl, combine pistachios, dark chocolate chips, and chopped dried apricots.
2. Sprinkle with sea salt and mix. Allow it to cool before storing.

Sesame-Ginger Seed Blend

Ingredients:

1/2 cup sesame seeds

1/2 cup sunflower seeds

1 tablespoon tamari or soy sauce

1 teaspoon grated ginger

Instructions:

1. In a dry pan, toast sesame seeds and sunflower seeds until golden.
2. Stir in tamari and grated ginger, coating the seeds evenly. Let it cool before enjoying.

Fiber-Rich Flaxseed and Coconut Mix

Ingredients

1/2 cup flaxseeds

1/2 cup shredded coconut

1/4 cup dried pineapple, diced

1/4 teaspoon vanilla extract

Instructions:

1. Combine flaxseeds, shredded coconut, dried pineapple, and vanilla extract in a bowl.
2. Blend thoroughly and preserve in a sealed receptacle.

Cacao and Hazelnut Energy Boost:

Ingredients:

1 cup hazelnuts

2 tablespoons cacao nibs

1 tablespoon honey

1/4 teaspoon sea salt

Instructions:

1. Roast hazelnuts in the oven at 350°F (175°C) for 10 minutes.
2. In a bowl, toss roasted hazelnuts with cacao nibs, honey, and sea salt. Let it cool before serving.

Turmeric and Pumpkin Spice Crunch

Ingredients:

1 cup mixed nuts (almonds, cashews, pecans)

1 tablespoon pumpkin spice mix

1 tablespoon coconut oil, melted

1 tablespoon maple syrup

Instructions:

1. In a bowl, mix mixed nuts with pumpkin spice mix.
2. Drizzle with melted coconut oil and maple syrup.
3. Spread on a baking sheet and bake at 325°F (163°C) for 12-15 minutes.
4. Let it cool before storing.

Gut Reset - Cleansing and Detox Recipes

Cleansing Soups and Broths

Healing Bone Broth

Ingredients:

2 lbs beef or chicken bones

1 onion, quartered

2 carrots, chopped

2 celery stalks, chopped

4 cloves garlic, smashed

1 tablespoon apple cider vinegar

Salt and pepper to taste

Instructions:

1. Roast bones in the oven at 400°F (200°C) for 30 minutes.
2. Place bones in a large pot, add vegetables, garlic, and vinegar.
3. Cover with water, bring to a boil, and then simmer for 12-24 hours.
4. Strain and season with salt and pepper.

Gut-Healing Miso Soup

Ingredients:

4 cups vegetable broth

2 tablespoons miso paste

1 cup shiitake mushrooms, sliced

1 cup bok choy, chopped

1 tablespoon tamari

1 teaspoon grated ginger

Instructions:

1. In a pot, heat vegetable broth until simmering.
2. In a bowl, mix miso paste with a small amount of broth until smooth.
3. Add miso mixture, mushrooms, bok choy, tamari, and ginger to the pot. Simmer for 10 minutes.

Turmeric and Lentil Detox Soup

Ingredients:

1 cup red lentils, rinsed

1 onion, diced

2 carrots, sliced

1 tablespoon turmeric

1 teaspoon cumin

6 cups vegetable broth

Salt and pepper to taste

Instructions:

1. In a pot, sauté onions and carrots until softened.
2. Add lentils, turmeric, cumin, and vegetable broth.
3. After bringing to a boil, cook the lentils until they are soft.
4. Season with salt and pepper.

Detoxifying Seaweed and Vegetable Soup:
Ingredients:

4 cups vegetable broth

1 cup wakame seaweed, soaked and chopped

1 cup daikon radish, sliced

1 cup kale, chopped

1 tablespoon rice vinegar

1 teaspoon sesame oil

Instructions:

1. Simmer the veggie stock in a saucepan.
2. Add seaweed, daikon radish, kale, rice vinegar, and sesame oil. Simmer for 15 minutes.

Cleansing Green Vegetable Soup

Ingredients

4 cups spinach, chopped

2 zucchinis, diced

1 leek, sliced

4 cups vegetable broth

1 tablespoon olive oil

1 teaspoon dried thyme

Instructions:

1. In a pot, sauté leeks in olive oil until softened.
2. Add zucchini, spinach, vegetable broth, and thyme. Simmer until vegetables are tender.

Lemon Ginger Detox Broth

Ingredients:

4 cups water

1 lemon, sliced

1-inch ginger, sliced

1 tablespoon apple cider vinegar

1 tablespoon fresh mint leaves

Cayenne pepper to taste

Instructions:

1. In a pot, combine water, lemon slices, ginger, apple cider vinegar, and mint leaves.
2. Bring to a simmer and let it steep for 10 minutes.
3. Add cayenne pepper for a kick.

Cabbage and Turmeric Cleansing Soup

Ingredients:

1 small head cabbage, shredded

2 carrots, grated

1 onion, diced

1 tablespoon turmeric

6 cups vegetable broth

Salt and pepper to taste

Instructions:

1. In a pot, sauté onions until translucent.
2. Add cabbage and carrots, sauté for 5 minutes.
3. Stir in turmeric and vegetable broth.
4. Simmer until vegetables are tender.
5. Season with salt and pepper.

Anti-Inflammatory Sweet Potato Soup

Ingredients:

2 sweet potatoes, peeled and diced

1 apple, peeled and chopped

1 onion, diced

1 teaspoon turmeric

4 cups vegetable broth

Salt and pepper to taste

Instructions:

1. In a pot, sauté onions until softened.
2. Add sweet potatoes and apple, sauté for 5 minutes.
3. Stir in turmeric and vegetable broth.

4. Simmer until sweet potatoes are tender.
5. Season with salt and pepper.

Beet and Carrot Detox Soup

Ingredients:

2 beets, peeled and diced

3 carrots, sliced

1 onion, diced

1 tablespoon ginger, grated

6 cups vegetable broth

Fresh parsley for garnish

Instructions:

1. In a pot, sauté onions and ginger until fragrant.
2. Add beets and carrots, sauté for 5 minutes.
3. Pour in vegetable broth and simmer until vegetables are tender.
4. Garnish with fresh parsley.

Cilantro and Avocado Green Soup

Ingredients

1 avocado, peeled and diced

1 cup fresh cilantro, chopped

1 cucumber, diced

2 cups spinach

1 lime, juiced

4 cups vegetable broth

Instructions:

1. In a blender, combine avocado, cilantro, cucumber, spinach, lime juice, and vegetable broth.
2. Blend until smooth. Serve chilled for a refreshing green detox soup.

Gut-Supporting Detox Juices

Green Gut Revitalizer

Ingredients:

2 cups kale

1 cucumber

1 green apple

1/2 lemon, peeled

1-inch ginger

Instructions:

1. Wash all ingredients thoroughly.
2. Juice kale, cucumber, green apple, lemon, and ginger.
3. Stir well and enjoy this refreshing green juice.

Citrus Digestive Tonic

Ingredients:

2 oranges, peeled

1 grapefruit, peeled

1 lime, peeled

1 tablespoon fresh mint leaves

Instructions:

1. Juice oranges, grapefruit, and lime.
2. Add fresh mint leaves to the juice.
3. Mix well and savor this citrusy digestive tonic.

Pineapple Ginger Soother

Ingredients:

2 cups pineapple chunks

1 apple

1-inch ginger

1 tablespoon aloe vera juice

Instructions:

1. Juice pineapple, apple, and ginger.
2. Stir in aloe vera juice for extra gut-soothing benefits.
3. Chill and enjoy this tropical, gut-supporting juice.

Carrot Turmeric Elixir

Ingredients:

4 carrots

1 orange, peeled

1-inch turmeric root

1/2 lemon, peeled

Instructions:

1. Juice carrots, orange, turmeric, and lemon.
2. Mix well and embrace the anti-inflammatory properties of turmeric.

Beetroot Detox Refresher

Ingredients:

2 beets, peeled

1 apple

1 cucumber

1 tablespoon chia seeds (soaked)

Instructions:

1. Juice beets, apple, and cucumber.
2. Add soaked chia seeds for an extra fiber boost.
3. Stir and enjoy this vibrant beetroot detox refresher.

Ginger Pear Cleanser

Ingredients

2 pears

1 cucumber

1-inch ginger

1/2 lemon, peeled

Instructions:

Juice pears, cucumber, ginger, and lemon.

Sip on this ginger pear cleanser for a gentle digestive boost.

Minty Melon Infusion

Ingredients:

2 cups watermelon cubes

1 cup cantaloupe cubes

1 tablespoon fresh mint leaves

Instructions:

1. Juice watermelon and cantaloupe.
2. Add fresh mint leaves for a burst of flavor.

3. Mix well and enjoy this hydrating and minty melon infusion.

Cabbage Apple Digestive Tonic

Ingredients:

1/2 small green cabbage

2 green apples

1 cucumber

1-inch ginger

Instructions:

1. Juice cabbage, green apples, cucumber, and ginger.
2. Stir well and experience the digestive benefits of cabbage.

Kiwi Berry Gut Soother

Ingredients:

4 kiwis, peeled

1 cup strawberries

1 cup blueberries

1 tablespoon flaxseeds (ground)

Instructions:

1. Juice kiwis, strawberries, and blueberries.
2. Add ground flaxseeds for added fiber and omega-3 fatty acids.
3. Mix and enjoy this antioxidant-rich gut soother.

Probiotic-rich Detox Smoothies

Berry Probiotic Bliss

Ingredients:

1 cup mixed berries (blueberries, strawberries, raspberries)

1 cup kefir

1 banana

1 tablespoon chia seeds

Ice cubes

Instructions:

1. Blend mixed berries, kefir, banana, and chia seeds until smooth.

2. Add ice cubes and blend again for a refreshing berry blast.

Green Probiotic Powerhouse

Ingredients:

2 cups spinach

1 cup pineapple chunks

1/2 cucumber

1 cup coconut water kefir

1 tablespoon flaxseeds (ground)

Instructions:

1. Blend spinach, pineapple, cucumber, coconut water kefir, and flaxseeds until smooth.
2. Enjoy this nutrient-packed green detox smoothie.

Mango Kombucha Delight

Ingredients:

1 cup mango chunks

1/2 cup Greek yogurt

1 cup kombucha (flavor of your choice)

1 tablespoon honey

Instructions:

1. Blend mango, Greek yogurt, kombucha, and honey until creamy.
2. Sip on this tropical and probiotic-rich delight.

Probiotic Avocado Berry Burst

Ingredients:

1/2 avocado

1 cup mixed berries (strawberries, blueberries, raspberries)

1 cup coconut milk kefir

1 tablespoon hemp seeds

Instructions:

1. Blend avocado, mixed berries, coconut milk kefir, and hemp seeds until silky smooth.
2. Revel in the creamy texture and fruity goodness.

Pineapple Ginger Kefir Cooler

Ingredients:

1 cup pineapple chunks

1 cup kefir

1-inch ginger

1 tablespoon chia seeds

Instructions:

1. Blend pineapple, kefir, ginger, and chia seeds until well combined.
2. Refresh yourself with this tangy and probiotic-rich cooler.

Papaya Probiotic Refresher

Ingredients:

1 cup papaya chunks

1 cup coconut water kefir

1 banana

1 tablespoon flaxseeds (ground)

Instructions:

1. Blend papaya, coconut water kefir, banana, and flaxseeds until smooth.
2. Savor the tropical goodness of this probiotic refresher.

Blueberry Yogurt Probiotic Smoothie Bowl

Ingredients:

1 cup blueberries

1/2 cup probiotic-rich yogurt

1 banana

2 tablespoons granola

1 tablespoon almond butter

Instructions:

1. Blend blueberries, yogurt, and banana until smooth.
2. Pour into a bowl and top with granola and almond butter for a satisfying smoothie bowl.

Turmeric Mango Lassi

Ingredients:

1 cup mango chunks

1/2 cup probiotic-rich yogurt

1/2 teaspoon turmeric

1 tablespoon honey

Instructions:

1. Blend mango, yogurt, turmeric, and honey until creamy.
2. Sip on this golden and probiotic-packed lassi.

Strawberry Basil Kombucha Smoothie

Ingredients:

1 cup strawberries

1 cup kombucha (flavor of your choice)

1/2 cup probiotic-rich yogurt

Fresh basil leaves

Instructions:

1. Blend strawberries, kombucha, yogurt, and fresh basil until well combined.
2. Enjoy the unique blend of strawberry sweetness and basil freshness.

Kiwi Probiotic Green Smoothie

Ingredients:

2 kiwis, peeled

1 cup spinach

1/2 cup Greek yogurt

1 tablespoon chia seeds

Instructions:

1. Blend kiwis, spinach, Greek yogurt, and chia seeds until smooth.
2. Revel in the vibrant green goodness of this probiotic-rich smoothie.

Peach Mint Kefir Cooler

Ingredients:

1 cup peaches, sliced

1 cup kefir

Fresh mint leaves

1 tablespoon honey

Instructions:

1. Blend peaches, kefir, mint leaves, and honey until well combined.
2. Sip on this peachy and refreshing kefir cooler.

Special Diets and Gut Health

Gut-Healthy Recipes for Vegetarians

Quinoa and Vegetable Stir-Fry

Ingredients:

1 cup quinoa, cooked

2 cups mixed vegetables (broccoli, bell peppers, carrots)

1 tablespoon olive oil

2 cloves garlic, minced

1 tablespoon soy sauce

1 teaspoon ginger, grated

1 tablespoon sesame seeds (optional)

Instructions:

1. In a pan, heat olive oil and sauté garlic and ginger.
2. Stir-fry the mixed vegetables until they become crisp-tender.
3. Stir in cooked quinoa and soy sauce, toss well.
4. Garnish with sesame seeds and serve.

Lentil and Spinach Soup

Ingredients:

1 cup green or brown lentils, soaked

1 onion, chopped

2 carrots, diced

2 cloves garlic, minced

4 cups vegetable broth

2 cups fresh spinach

1 teaspoon cumin

Salt and pepper to taste

Instructions:

1. In a pot, sauté onions and garlic until translucent.
2. Add lentils, carrots, cumin, and vegetable broth. Simmer until lentils are tender.
3. Stir in fresh spinach and cook until wilted. Season with salt and pepper.

Chickpea and Avocado Salad

Ingredients:

1 can chickpeas, drained and rinsed

1 avocado, diced

1 cucumber, chopped

1 cup cherry tomatoes, halved

1/4 cup red onion, finely chopped

2 tablespoons olive oil

1 tablespoon lemon juice

Fresh cilantro or parsley for garnish

Instructions:

1. In a large bowl, combine chickpeas, avocado, cucumber, cherry tomatoes, and red onion.
2. Drizzle olive oil and lemon juice over the salad. Toss gently.
3. Add some cilantro or fresh parsley as a garnish before serving.

Roasted Brussels Sprouts with Balsamic Glaze
Ingredients:

1 lb Brussels sprouts, halved

2 tablespoons olive oil

Salt and pepper to taste

2 tablespoons balsamic glaze

1/4 cup chopped walnuts (optional)

Instructions:

1. Add salt, pepper, and olive oil to Brussels sprouts and toss.
2. Roast in the oven at 400°F (200°C) until golden brown.
3. Drizzle with balsamic glaze and sprinkle with chopped walnuts if desired.

Sweet Potato and Black Bean Bowl

Ingredients:

2 sweet potatoes, diced

1 can black beans, drained and rinsed

1 cup corn kernels

1 teaspoon cumin

1 teaspoon chili powder

1 avocado, sliced

Fresh lime wedges for serving

Instructions:

1. Bake sweet potatoes until they become soft.

2. In a pan, heat black beans, corn, cumin, and chili powder.
3. Assemble bowls with sweet potatoes, black bean mixture, and avocado slices. Serve with lime wedges.

Spinach and Feta Stuffed Mushrooms

Ingredients:

12 large mushrooms, stems removed

2 cups fresh spinach, chopped

1/2 cup feta cheese, crumbled

2 cloves garlic, minced

2 tablespoons olive oil

Salt and pepper to taste

Instructions:

1. Preheat the oven to 375°F (190°C).
2. In a pan, sauté garlic in olive oil, then add spinach and cook until wilted.
3. Mix spinach with feta cheese, salt, and pepper.
4. Stuff mushrooms with the spinach and feta mixture.
5. Bake for 15-20 minutes until mushrooms are tender.

Mediterranean Quinoa Salad

Ingredients:

1 cup quinoa, cooked

1 cup cherry tomatoes, halved

1 cucumber, diced

1/2 cup Kalamata olives, sliced

1/2 cup feta cheese, crumbled

2 tablespoons olive oil

1 tablespoon red wine vinegar

Fresh basil or oregano for garnish

Instructions:

1. In a large bowl, combine quinoa, cherry tomatoes, cucumber, olives, and feta.
2. Drizzle with olive oil and red wine vinegar. Toss gently.
3. Before serving, garnish with fresh oregano or basil.

Cauliflower and Chickpea Curry

Ingredients:

1 cauliflower, cut into florets

1 can chickpeas, drained and rinsed

1 onion, chopped

2 cloves garlic, minced

1 can coconut milk

2 tablespoons curry powder

Salt and pepper to taste

Instructions:

1. In a pan, sauté onions and garlic until softened.
2. Add cauliflower, chickpeas, curry powder, coconut milk, salt, and pepper.

Grilled Eggplant and Quinoa Stacks

Ingredients:

1 large eggplant, sliced

1 cup quinoa, cooked

1 tomato, sliced

1/2 cup fresh mozzarella, sliced

Fresh basil leaves

Balsamic glaze for drizzling

Olive oil for brushing

Salt and pepper to taste

Instructions:

1. Grill eggplant slices until they are soft, brushing them with olive oil.
2. Assemble stacks by layering grilled eggplant, quinoa, tomato, and mozzarella.
3. Top with fresh basil leaves, drizzle with balsamic glaze, and season with salt and pepper.

Zucchini Noodles with Pesto

Ingredients:

4 medium zucchinis, spiralized

1 cup cherry tomatoes, halved

1/4 cup pine nuts, toasted

1/2 cup fresh basil leaves

1/2 cup grated Parmesan cheese

1/2 cup extra virgin olive oil

2 cloves garlic

Salt and pepper to taste

Instructions:

1. In a blender, combine basil, pine nuts, Parmesan, garlic, and olive oil. Blend until smooth to make pesto.
2. Toss zucchini noodles with pesto, cherry tomatoes, salt, and pepper.
3. Serve garnished with additional Parmesan if desired.

Roasted Butternut Squash and Kale Salad

Ingredients:

1 small butternut squash, peeled and cubed

4 cups kale, chopped

1/2 cup dried cranberries

1/4 cup pumpkin seeds, toasted

2 tablespoons balsamic vinaigrette

Salt and pepper to taste

Instructions:

1. Roast butternut squash in the oven until golden and tender.
2. Massage kale with balsamic vinaigrette to soften.

3. Combine roasted butternut squash, kale, dried cranberries, and pumpkin seeds. Season with salt and pepper.

Black beans and Quinoa Stuffed Bell Peppers

Ingredients:

4 bell peppers, halved and seeds removed

1 cup quinoa, cooked

1 can black beans, drained and rinsed

1 cup corn kernels

1 cup salsa

1 teaspoon cumin

1 teaspoon chili powder

1 cup shredded cheddar cheese

Instructions:

1. Preheat the oven to 375°F (190°C).
2. Quinoa, black beans, corn, salsa, cumin, and chili powder should all be combined in a bowl.
3. Stuff bell peppers with the quinoa mixture and top with shredded cheddar cheese.
4. Bake until the cheese melts and the peppers are soft.

Gut-Friendly Options for Gluten-Free Diets

Quinoa and Vegetable Buddha Bowl

Ingredients:

1 cup cooked quinoa

1 cup roasted sweet potatoes

1 cup sautéed kale

1/2 cup shredded carrots

1/4 cup sliced avocado

2 tablespoons gluten-free tamari or soy sauce

1 tablespoon olive oil

Instructions:

1. Arrange quinoa, sweet potatoes, kale, carrots, and avocado in a bowl.
2. Drizzle with tamari and olive oil.
3. Toss gently and enjoy this nourishing gluten-free Buddha bowl.

Salmon and Asparagus Quinoa Salad

Ingredients:

1 cup cooked quinoa

6 oz grilled salmon, flaked

1 cup steamed asparagus, chopped

1/4 cup cherry tomatoes, halved

2 tablespoons lemon vinaigrette (olive oil, lemon juice, Dijon mustard)

Salt and pepper to taste

Instructions:

1. Combine quinoa, grilled salmon, asparagus, and cherry tomatoes in a bowl.
2. Drizzle with lemon vinaigrette and season with salt and pepper.
3. Toss gently and serve as a refreshing gluten-free salad.

Zucchini Noodles with Pesto and Cherry Tomatoes

Ingredients:

4 medium zucchinis, spiralized

1/2 cup homemade or store-bought pesto (ensure it's gluten-free)

1 cup cherry tomatoes, halved

1/4 cup pine nuts, toasted

Salt and pepper to taste

Instructions:

1. Pesto-coated zucchini noodles should be thoroughly mixed.
2. Add cherry tomatoes and toasted pine nuts.
3. Season with salt and pepper and serve this light and flavorful gluten-free dish.

Grilled Chicken and Quinoa Stuffed Bell Peppers

Ingredients:

4 bell peppers, halved and seeds removed

1 cup cooked quinoa

1 cup grilled chicken, diced

1/2 cup black beans, drained and rinsed

1/2 cup corn kernels

1/2 cup salsa

1 teaspoon cumin

1 teaspoon chili powder

1 cup shredded cheese (ensure it's gluten-free)

Instructions:

1. Preheat the oven to 375°F (190°C).
2. In a bowl, mix quinoa, grilled chicken, black beans, corn, salsa, cumin, and chili powder.
3. Stuff bell peppers with the quinoa mixture and top with shredded cheese.
4. Bake until the cheese melts and the peppers are soft.

Eggplant and Tomato Gratin

Ingredients:

1 large eggplant, sliced

2 cups cherry tomatoes, halved

1/2 cup gluten-free breadcrumbs

1/4 cup grated Parmesan cheese

2 tablespoons olive oil

2 cloves garlic, minced

Fresh basil for garnish

Instructions:

1. Preheat the oven to 400°F (200°C).
2. Layer sliced eggplant and cherry tomatoes in a baking dish.
3. In a bowl, combine gluten-free breadcrumbs, Parmesan, olive oil, and minced garlic.
4. Sprinkle the breadcrumb mixture over the vegetables.
5. Bake until the top is golden brown. Garnish with fresh basil before serving.

Shrimp and Vegetable Stir-Fry with Cauliflower Rice

Ingredients:

1 lb shrimp, peeled and deveined

2 cups broccoli florets

1 red bell pepper, sliced

1 cup snap peas

2 tablespoons gluten-free tamari or soy sauce

1 tablespoon sesame oil

4 cups cauliflower rice (pre-made or grated)

2 cloves garlic, minced

Instructions:

1. In a wok or large pan, stir-fry shrimp, broccoli, bell pepper, and snap peas with tamari and sesame oil.
2. In a separate pan, sauté cauliflower rice with minced garlic until tender.
3. Serve the shrimp and vegetables over cauliflower rice for a gluten-free stir-fry.

Mediterranean Stuffed Portobello Mushrooms

Ingredients:

4 large portobello mushrooms, stems removed

1 cup quinoa, cooked

1/2 cup cherry tomatoes, diced

1/4 cup Kalamata olives, chopped

1/4 cup feta cheese, crumbled

2 tablespoons olive oil

Fresh parsley for garnish

Instructions:

1. Preheat the oven to 375°F (190°C).
2. In a bowl, mix quinoa, cherry tomatoes, Kalamata olives, and feta cheese.

3. Spoon the quinoa mixture into each portobello mushroom.
4. Drizzle with olive oil and bake until the mushrooms are tender.
5. Garnish with fresh parsley before serving.

Coconut-Curry Chickpea Stew

Ingredients:

2 cans chickpeas, drained and rinsed

1 can coconut milk

1 cup sweet potatoes, diced

1 cup spinach

1 onion, chopped

2 cloves garlic, minced

2 tablespoons gluten-free curry powder

1 tablespoon olive oil

Salt and pepper to taste

Instructions:

1. In a pot, sauté onions and garlic in olive oil until softened.

2. Add sweet potatoes, chickpeas, coconut milk, and curry powder. Simmer until sweet potatoes are tender.
3. Stir in spinach and cook until wilted. Season with salt and pepper.

Turkey and Vegetable Lettuce Wraps

Ingredients:

1 lb ground turkey

1 cup mixed bell peppers, diced

1 cup carrots, julienned

1/2 cup water chestnuts, chopped

1/4 cup gluten-free hoisin sauce

2 tablespoons gluten-free soy sauce

1 tablespoon sesame oil

Iceberg lettuce leaves for wrapping

Green onions for garnish

Instructions:

1. In a pan, cook ground turkey until browned.
2. Add bell peppers, carrots, and water chestnuts. Add the sesame oil, hoisin sauce, and soy sauce and stir.

3. Spoon the turkey and vegetable mixture into lettuce leaves. Garnish with green onions.

Spinach and Artichoke Quinoa Casserole

Ingredients:

1 cup quinoa, cooked

1 cup fresh spinach, chopped

1 can artichoke hearts, drained and chopped

1 cup shredded mozzarella cheese (ensure it's gluten-free)

1/2 cup grated Parmesan cheese

1/2 cup gluten-free breadcrumbs

1/2 cup Greek yogurt

2 cloves garlic, minced

Salt and pepper to taste

Instructions:

1. Preheat the oven to 375°F (190°C).
2. In a bowl, combine quinoa, spinach, artichoke hearts, mozzarella, Parmesan, breadcrumbs, Greek yogurt, garlic, salt, and pepper.
3. Transfer the mixture to a baking dish and bake until the top is golden and bubbly.

Cauliflower Pizza Crust with Veggie Toppings

Ingredients:

1 head cauliflower, riced

1/2 cup almond flour

2 eggs

1 teaspoon dried oregano

1/2 teaspoon garlic powder

1/4 teaspoon salt

1/4 teaspoon black pepper

Pizza sauce (gluten-free)

Mozzarella cheese (ensure it's gluten-free)

Assorted vegetable toppings (bell peppers, tomatoes, olives, etc.)

Fresh basil for garnish

Instructions:

1. Preheat the oven to 425°F (220°C).
2. Mix riced cauliflower, almond flour, eggs, oregano, garlic powder, salt, and pepper to form a dough.
3. Press the dough onto a parchment-lined baking sheet to create a pizza crust.
4. Bake the crust for firmness and a rich brown color.

5. Spread gluten-free pizza sauce, sprinkle mozzarella, and add vegetable toppings.
6. Bake until the cheese is melted and bubbly. Garnish with fresh basil before serving.

Conclusion

In the culinary journey through "The Gut Check cookbook: 100+ delicious recipes you'll want to eat to help you Unleash the Power of Your Microbiome to Reverse Disease and heal your gut" we've explored the intricate world of our gut microbiome and its profound impact on our overall well-being. Cherri J. Diaz has provided us with a comprehensive guide to understanding, nourishing, and revitalizing our gut health, emphasizing the transformative potential it holds for our mental, physical, and emotional health.

Embracing the Microbial Symphony

As we delved into the pages of this cookbook, the microbial symphony within our bodies became increasingly clear. Our gut is not merely a passive player in our health but a dynamic and intelligent ecosystem orchestrating the complex dance of trillions of microorganisms. The revelations shared by Cherri J. Diaz reinforce the age-old wisdom attributed to Hippocrates, affirming that indeed, "all disease begins in the gut." However, "Gut Check" goes beyond identifying the problem; it provides a roadmap for healing and rejuvenation.

A Culinary Journey of Healing

The recipes presented throughout the cookbook are more than just a collection of delicious meals. They represent a commitment to nurturing and supporting our microbiome, recognizing it as the epicenter of our vitality. From breakfast to dinner, and every snack in between, each dish is crafted with the intention of promoting gut health and, by extension, fostering holistic well-being.

The breakfast offerings, ranging from gut-nourishing smoothies to fiber-rich quinoa bowls, set the tone for a day filled with nutritious choices. Lunch options emphasize the inclusion of lean proteins, vibrant vegetables, and probiotic-rich ingredients, providing sustained energy and supporting digestive wellness. Moving on to dinner, the array of recipes showcases the creative integration of lean proteins with probiotic marinades, ensuring a delightful and healthful dining experience.

The salads presented in this cookbook are not mere side dishes but wholesome creations that marry fiber-rich greens with probiotic dressings, turning salads into a feast for both the taste buds and the microbiome. Meanwhile, the snacks and drinks section elevates these often-overlooked elements to essential components of our daily gut-health routine, offering alternatives that are as satisfying as they are beneficial.

Beyond Taste: The Healing Potential

As we've navigated through the cookbook, it's crucial to recognize that these recipes extend beyond culinary delights; they are a form of self-care and a pathway to healing. The choices we make in our kitchens have the power to influence our immune systems, hormone levels, mental health, and even our longevity. "Gut Check" doesn't just stop at presenting the science; it empowers us to take charge of our health by treating our microbes right.

The emphasis on fermented foods, rich in probiotics, highlights their transformative potential. From gut-healing soups and stews to fiber-rich salads and probiotic-infused sides, the recipes provided cater to diverse tastes and preferences. The one-pot meals for easy digestion and the array of probiotic snacks showcase the versatility of a gut-friendly diet, proving that taking care of our microbiome need not be a monotonous task but a flavorful and enjoyable journey.

A Call to Action: Your Gut, Your Health

As we reach the conclusion of this culinary expedition, it's essential to acknowledge that the power to transform our health lies within our hands—quite literally, in our kitchens. Dr. Gundry's expertise and the insights shared in "Gut Check" empower us to make informed choices that resonate with our bodies' unique needs. It's a call to action, urging us to be mindful of the symbiotic relationship we share with our microbiome.

The principles woven into this cookbook are not meant to be fleeting trends but enduring lifestyle choices. They invite us to reevaluate our relationship with food, viewing it not just as sustenance but as a tool for nurturing the trillions of microorganisms that call our bodies home. The recipes are an invitation to savor not only the flavors but the potential for transformation with each bite.

The Future of Health: A Gut-First Approach

In an era where health trends come and go, the gut-first approach advocated in "Gut Check" stands as a timeless pillar of well-being. It's a shift from quick fixes to sustainable practices, from external remedies to internal rejuvenation. The cookbook encourages us to think of our gut as a garden, one that requires tending, nourishing, and occasional pruning to flourish.

As you embark on your gut-health journey, remember that it's not about perfection but progress. Every conscious choice you make in your kitchen contributes to the resilience and vibrancy of your microbiome. It's an investment in your health that pays dividends in vitality, clarity, and overall wellness.

Gratitude and Continued Wellness

In closing, we express our gratitude for joining us on this voyage through the pages of "Gut Check." It's our sincere hope that this cookbook serves as a trusted companion on your path to health and vitality. May the recipes inspire creativity in your kitchen, the insights deepen your understanding of your body, and the journey bring about positive transformation.

As you savor the flavors and nourish your gut, remember that you're not just eating; you're engaging in an act of self-love and self-care. Your gut, with its diverse microbial community, is your partner in health, and by embracing its well-being, you're nurturing your own.

Here's to your continued journey of vibrant health, one gut-friendly meal at a time. Cheers to the power of your microbiome and the limitless potential it holds for a healthier, happier you.

Our team of editors and writers spent a lot of time working on this cookbook to get the best for you. Did this book help you in some way?

If so, I'd love to hear about it. Honest reviews help readers find the right book for their needs. Please help more readers make better decisions by sharing your thoughts on this cookbook.

30 Days Meal Plan

WEEKLY *Meal* PLAN

MONTH
WEEK

Shopping
LIST

	BREAKFAST	LUNCH	DINNER
MONDAY			
TUESDAY			
WEDNESDAY			
THURSDAY			
FRIDAY			
SATURDAY			
SUNDAY			

WEEKLY *Meal* PLAN

Shopping LIST

	BREAKFAST	LUNCH	DINNER
MONDAY			
TUESDAY			
WEDNESDAY			
THURSDAY			
FRIDAY			
SATURDAY			
SUNDAY			

WEEKLY *Meal* PLAN

MONTH:
WEEK:

Shopping
LIST

	BREAKFAST	LUNCH	DINNER
MONDAY			
TUESDAY			
WEDNESDAY			
THURSDAY			
FRIDAY			
SATURDAY			
SUNDAY			

WEEKLY *Meal* PLAN

MONTH
WEEK

	BREAKFAST	LUNCH	DINNER
MONDAY			
TUESDAY			
WEDNESDAY			
THURSDAY			
FRIDAY			
SATURDAY			
SUNDAY			

WEEKLY *Meal* PLAN

MONTH:
WEEK:

Shopping LIST

	BREAKFAST	LUNCH	DINNER
MONDAY			
TUESDAY			
WEDNESDAY			
THURSDAY			
FRIDAY			
SATURDAY			
SUNDAY			

www.ingramcontent.com/pod-product-compliance
Lightning Source LLC
Chambersburg PA
CBHW071044290526
45795CB00004B/1312